Contents

Preface … 3

Foreword … 9

Chapter

1 Run of Rafters and Pitch of Roof … 13

2 Length of Rafters, Hips and Valleys … 17

3 Plumb and Seat Bevels of Rafters, Hips and Valleys … 25

4 Edge Bevels of Hips, Valleys and Jacks … 29

5 Marking and Cutting the Roof Timbers … 35

6 Marking and Cutting Jack Rafters and Purlins … 45

7 Assembling and Erecting the Roof Timbers … 51

8 Battening Roof for Slates … 55

Metric Roofing Ready Reckoner for every 2½° pitch from 17½° to 60° … 61

The Carpenter's Metric Roofing Ready Reckoner

and

Steel Square Roofing Explained and Simplified
combined in one volume

METRIC EDITION

BY

W. E. GRAY

A.M.Inst. B.E.

Stobart Davies
Ammanford

Copyright © 1972

Stobart Davies Ltd

All rights reserved. No part of this publication may be reproduced or transmitted in any form or by any means, electronic or mechanical, including photocopying, tape recording, or any information storage and retrieval system, without the publisher's permission in writing.

Metric Edition first published 1972,
Reprinted 2021 by Stobart Davies Ltd.
Pontyclerc, Penybanc Road,
Ammanford SA1B 3HP

ISBN 978-0-85442-004-9

Printed and bound by Dinefwr Press, Llandybie

Preface to first metric edition 1972

The Carpenter's Roofing Ready Reckoner was first published in 1948 and nearly 16,000 copies have now been sold, some as far afield as Arabia, The Sudan, Ghana, Trinidad, Nigeria and all the Commonwealth countries and Colonies. Hundreds of letters have been received from carpenters and builders from many parts of the world saying how helpful they have found this " . . . wonderful little book" which "saves endless time and money" and ". . . is the most easily understood" and " . . . the best method of roofing they have ever come across".

Because my Roofing Ready Reckoner has been so helpful to so many carpenters, I felt it was necessary to explain the principles of "Steel Square" roofing and show how lengths and bevels are found, hence my second book. Steel Square Roofing Explained and Simplified. By using this book a carpenter could find all lengths and bevels for himself and come to understand "Steel Square" roofing more fully.

Considering the number of excellent books that have been written on roofing with the

"Steel Square" it is surprising how few carpenters understand or use it. Even though its use is taught in technical schools, many fail to grasp it and those who do, unless they are using the square continually, quickly forget.

How many, if they did not use the square for a year or two, could take it up and without consulting their text books and revising their studies, correctly mark the timbers for a roof with hips, valleys, jacks and purlins? Very few I expect.

The demand for the Roofing Ready Reckoner during the past 23 years, continues and has fully justified my conviction of the need for this simplified method of "Steel Square" roofing which leaves nothing to memory.

Many carpenters, particularly the older ones, will be somewhat confused when the changeover to the metric system of measurement comes into operation. All local authority plans will be in metric by 1972 and private architects will no doubt follow suit. In view of this and also the tendency to specify roof pitches in degrees instead of inches of rise per foot of run, and in response to the many requests for help that I have received, I have rewritten my two books combining them into one volume, translating all tables and calculations into millimetres.

Translating the tables and drawings into the metric system has been no mean task, as the two systems are not compatible, and all calculations have had to be independently worked out without reference to the imperial system of measurement, but the principles of roofing with the "Steel Square" and the principles of the Roofing Ready Reckoner remain the same, and I hope this new Metric Roofing Ready Reckoner will give the same help to as many, if not more, carpenters as my previous books have done.

All the unsolicited testimonials quoted in this metric edition are from letters received from users of my Roofing Ready Reckoner.

APPRECIATION
I should like to acknowledge my indebtedness to my son-in-law W. T. Pearce for the excellence of the illustrations which he has re-drawn from my original sketches.

W. E. GRAY

Foreword

Roofing by the "Steel Square" method, is without doubt the quickest and most accurate known, but only a small percentage of carpenters use it, the majority finding it too complicated and the opportunities for practising its use, too few.

There has long been felt a need for a simpler method whereby the whole length of rafters, hips and valleys can be obtained instead of just the length for one foot of run; now the metric system is coming into practice, the need is even greater.

Many times have I been asked to explain the method I myself use and in writing the following chapters, explaining how the various lengths and bevels are found, and in compiling the Ready Reckoner tables in the metric system, I have endeavoured to present "Steel Square" roofing without its complications and to so simplify its use that all carpenters will be able to understand and use it.

Before studying the "Metric Ready Reckoner" tables it is essential to understand the meaning of the terms used and their relation to roofing. The illustrations are not to scale and some of the roof members are enlarged out of all proportion, the more to emphasize points it is desired to explain.

Because all imperial measure rules, tapes and roofing squares will gradually become obsolete after the changeover to metric, this new Metric Roofing Ready Reckoner has been written and compiled. It covers a wider range of roof pitches with tables giving lengths, bevels, differences in jacks etc., for every 2½° of pitch, from 17½° to 60°, all in millimetres; thus it can be used with any roofing square that has millimetre graduations.

Your purchase of this Metric Roofing Ready Reckoner is appreciated and I hope it will prove as useful and helpful to you, as my earlier books have done to the many thousands of carpenters who have used them.

W. E. GRAY

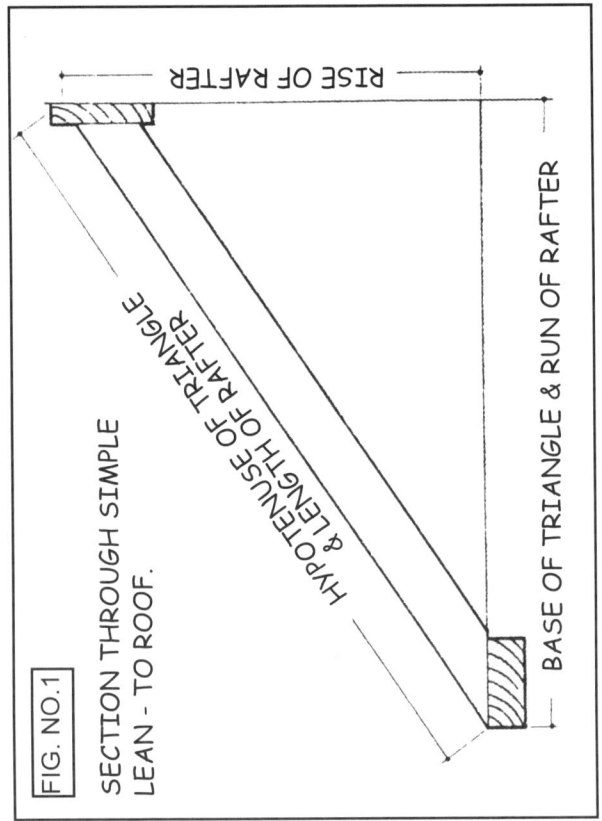

FIG. NO.1
SECTION THROUGH SIMPLE LEAN-TO ROOF.

CHAPTER 1

Run of Rafters and Pitch of Roof

All roofs are made up of a series of right angle triangles as will be seen from illustrations Nos. 1 and 2. In each illustration, the run of the rafter forms the base of the triangle. The rise of the rafter is the altitude or vertical side of the triangle, and the rafters themselves form the hypotenuse of the triangles.

The term "run of rafter" means the horizontal distance a rafter covers in attaining its required height; in other words, it is the distance from the outside edge of the wallplate to the face of the wall for a lean-to roof, and from the outside edge of the wall plate to a point directly below the centre of the ridge for an independent or span roof.

The "rise" of the rafter is the vertical height to which a rafter rises above its starting point, i.e. the wall plate. In other words, the vertical distance from the level of the wall plate to the highest point of the rafter; against the wall for a lean-to roof, and to the centre of the ridge for an independent or span roof.

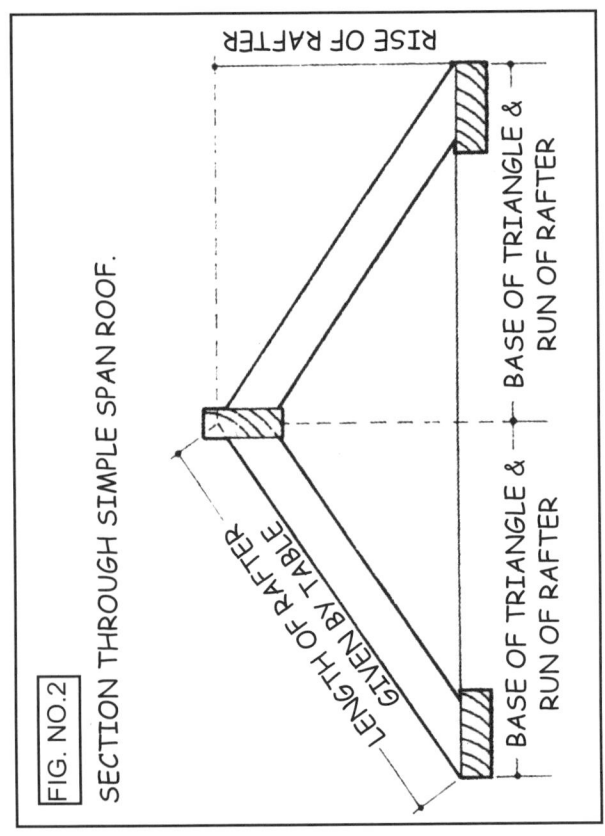

FIG. NO.2 SECTION THROUGH SIMPLE SPAN ROOF.

The pitch of the roof is the rise of the roof in relation to its span or width of the building ; it used to be referred to in fractions of the width of the building, such as one third pitch, five twelfths pitch, or half pitch, or in inches of rise per foot of run. Fig. 2.

Now that the Metric System is to be used, it is no longer practical to use those terms, particularly as architects generally specify roof pitches in degrees from the horizontal without reference to the width of the building, so "Rise per foot of run" will no longer be applicable. "Degrees of Pitch" taking its place.

Use 300 mm of run on the blade of the square, instead of 12" as before. with the rise in millimetres per 300 mm of run, on the tongue. The rise in millimetres per 300 mm of run for each given pitch has been calculated and is shown on each table; these given figures used on the tongue of the square in conjunction with 300 mm on the blade, give the plumb and seat cuts of the common rafters for each given degree of pitch, e.g. the figure to be used on the tongue of the square for the plumb and seat cuts of a 35° pitch rafter is 210 mm which is also the rise for 300 mm of its run.

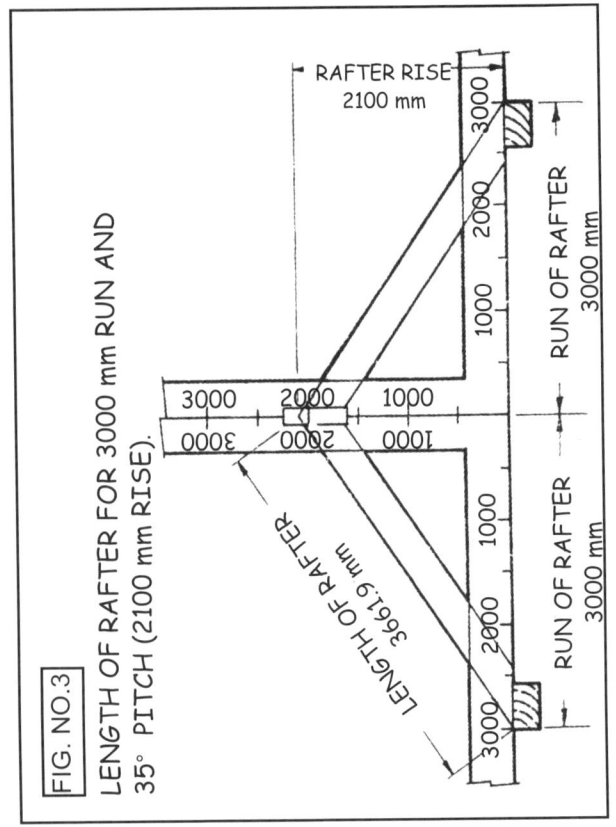

FIG. NO.3
LENGTH OF RAFTER FOR 3000 mm RUN AND 35° PITCH (2100 mm RISE).

CHAPTER 2

Length of Rafters, Hips and Valleys

Fig. 3 illustrates two giant squares, ten times bigger than normal, each 6,000 mm x 4,500 mm standing in position as they would be in the actual roof. The wall plates are fixed at the 3,000 mm mark, i.e. 6,000 mm out to out of wall plates, representing 6,000 mm total width of the building; thus each rafter has a run of 3,000 mm.

A 35° pitch roof has a rise of 2,100 mm in 3,000 mm run of rafter. The length of each rafter is the diagonal of the 3,000 mm x 2,100 mm rectangle, which measures 3,661·9 mm, or, dividing by ten, i.e. moving the decimal point one place to the left, the rafter length is 366.19 mm as given in the 35° table in The Carpenter's Metric Roofing Ready Reckoner .19 should be ignored and the length read as 366 mm.

To find the length of the hips and valleys, we must remember that on plan, they are the diagonal of the square formed by the run of

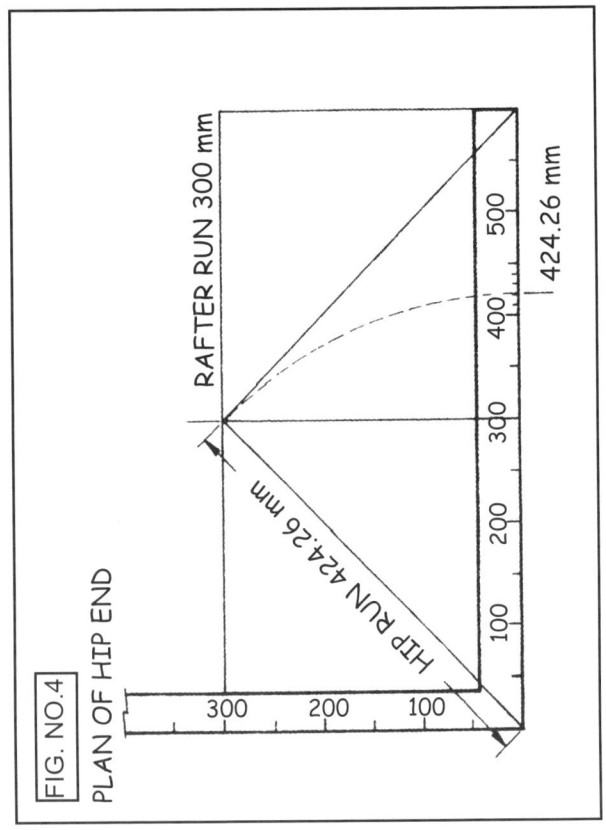

the common rafters, which we have reduced to 300 mm.

The diagonal of a 300 mm square is 424.26 mm which for all practical purposes is counted as 424 mm. As the rafters. hips and valleys all have the same total rise. the length of the hips and valleys is the diagonal of the rectangle formed by 424 mm x 210 mm, which is 473.39 mm (see Fig. 4). this would be read as 473 mm. Thus it will be seen that for every 300 mm of run of the common rafter, the hip and valley have a run of 424 mm.

This is the only occasion with this method of roofing, in which the figure of 300 mm is not used on the blade of the square.

In every right triangle the square of the hypotenuse is equal to the sum of the squares of the other two sides. To find the length of the hypotenuse of a right triangle, i.e. the length of a rafter with a run of 4 units and a rise of 3 units, square each of the dimensions and add them together thus, $4^2 = 16$ and $3^2 = 9$ and $16 + 9 = 25$ which is the square of the hypotenuse or rafter. The square root of 25 is 5, therefore the length of the hypotenuse or rafter, is 5 units.

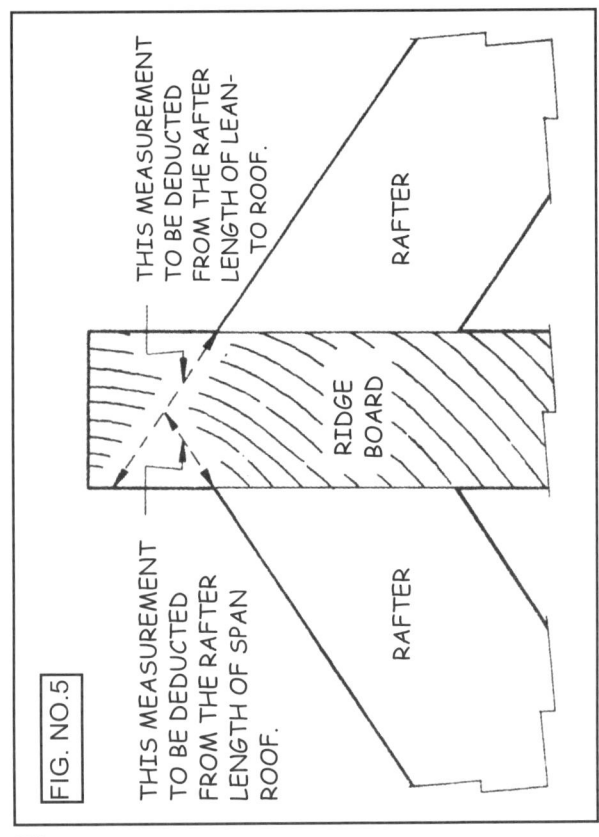

The same principle applies to all right triangles and all the tables of lengths of rafters, hips and valleys have been calculated in this way to the fifth decimal place, the last being discarded if below 5, or, if above 5, 1 has been added to the fourth decimal place, and all lengths read to the fourth decimal place. All lengths have also been checked on large scale drawings.

The lengths given in the tables are from the outside edge of the wall plate to the centre of the ridge, or to the face of the adjacent wall for a lean-to roof.

But, since a ridge board is introduced into the length of the run, the thickness of that board must be deducted from the length of the rafters as given in the tables of length; half the thickness from each rafter in the case of an independent or span roof (Fig. 2) and the whole thickness from the rafter of a lean-to roof (Fig.1).

As the measurement of the thickness of the ridge must be taken at the same angle as the pitch of the rafter, this measurement will vary accordingly. The simplest way of measuring that thickness is to mark the rafter plumb cut

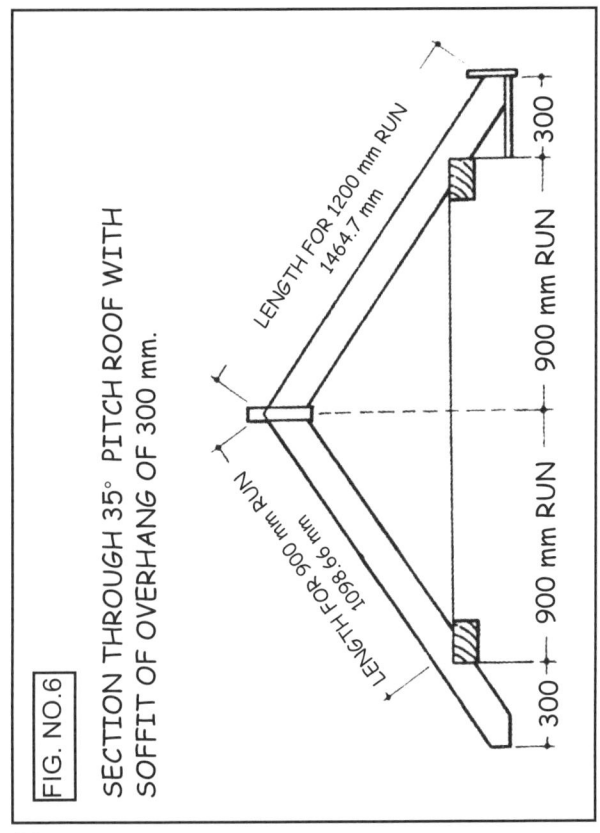

FIG. NO.6

SECTION THROUGH 35° PITCH ROOF WITH SOFFIT OF OVERHANG OF 300 mm.

on the edge of the ridge board, (Fig. 5) and measure the actual line; that measurement is the thickness that must be deducted from the length obtained from the tables.

Fig. 6 illustrates a 35° pitch independent or span roof with the rafters extending beyond the walls to form an overhang, or soffit and facia. The run of the rafter in this case is 900 mm plus the overhang or soffit which is 300 mm, making a total run of 1,200 mm. The rafter length for 1,000 mm run is 1,220.6 mm and for 200 mm is 244.13 mm which added together is 1,464.13 mm from which length the allowance for half the thickness of the ridge must be deducted as explained in the previous paragraph. It is preferable however, to take the run as from the outside edge of the wall plate i.e. 900 mm, and allow the over hang sufficiently long enough to be cut off to the required overhang after a 11 the rafters, hips and valleys are fixed, thus assuring a straight line for fixing facia board.

". . . that grand little book . . . The Carpenter 's Roofing Ready Reckoner . . . *most of the lads purchased this book years ago and they have done yeoman service . . . we are lost without them.*"

(H . H ., Whitby, Yorks.)

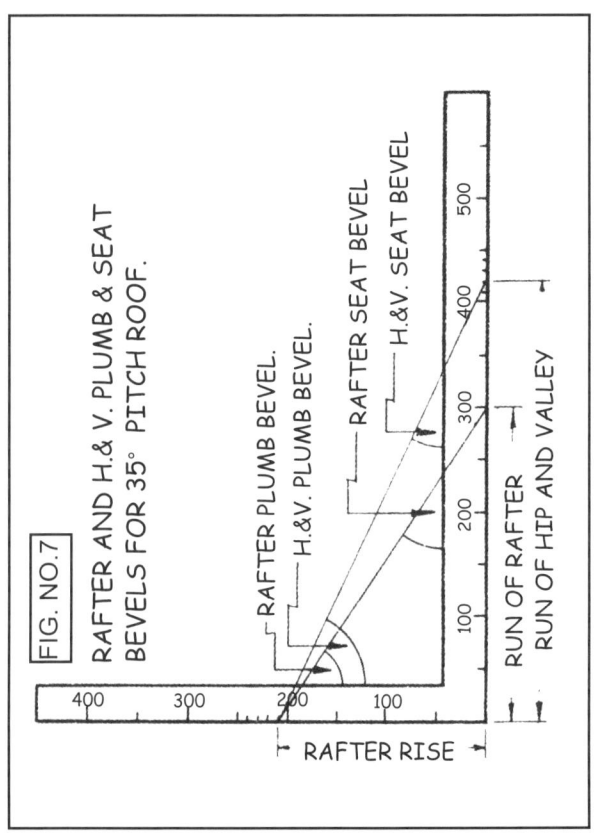

CHAPTER 3

Plumb and Seat Bevels of Rafters, Hips and Valleys

To find the plumb and seat cuts or bevels of rafters, hips and valleys, let us look at Fig. 7 which is a simple illustration of the square in a 35° pitch setting. The rafter is represented by a line drawn from the 300 mm mark on the blade i.e. the run of the rafter, to the 210 mm mark on the tongue of the square i.e. the rise of the rafter. The angle formed by the blade of the square and the rafter line, is the seat cut of the rafter, and the angle formed by the rafter line and the tongue of the square, is the plumb bevel of the rafter.

The hip and valley line runs from the 210 mm mark on the tongue, (as they and the rafters all have the same total rise), to the 424 mm mark on the blade of the square, which is the run of the hip and valley, i.e. the diagonal of a 300 mm square. The angle formed by the hip line and the blade of the square, is the seat bevel of the hips and valleys, and the angle between the hip line and the tongue of the

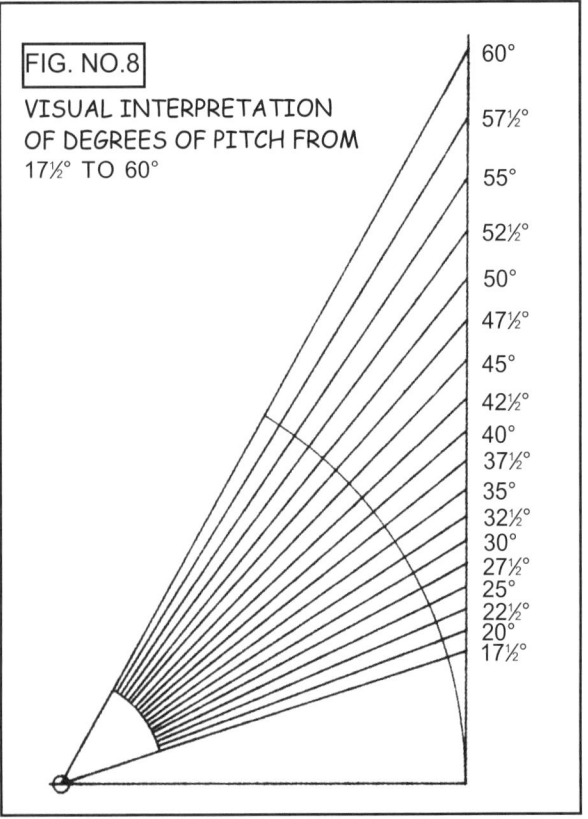

square, is the plumb bevel of the hips and valleys.

This is the only occasion, with this method of roofing, in which the figure 300 mm is not used on the blade of the square.

We have now found the lengths of rafters, hips and valleys by measuring the diagonals of their respective run and rise, and have found the plumb and seat bevels as formed by those diagonals against the blade and tongue of the square. Fig. 8 illustrates roof pitches from 17½° to 60° and presents a visual interpretation of degrees of pitch.

". . . copies of your Steel Square Roofing and Roofing Ready Reckoner and am delighted with their clarity and simplicity."

(R. S. H., Dorking, Surrey)

"I am still using your Roofing Book which is the best one I have come across."

(H. C. K ., St. Leonards-on-Sea, Sussex)

"Indeed it is possibly the best book on roofing ever written . . . I feel it is a booklet which will never be equalled, so it must never go out of print."

(D. A .. Brentwood, Essex)

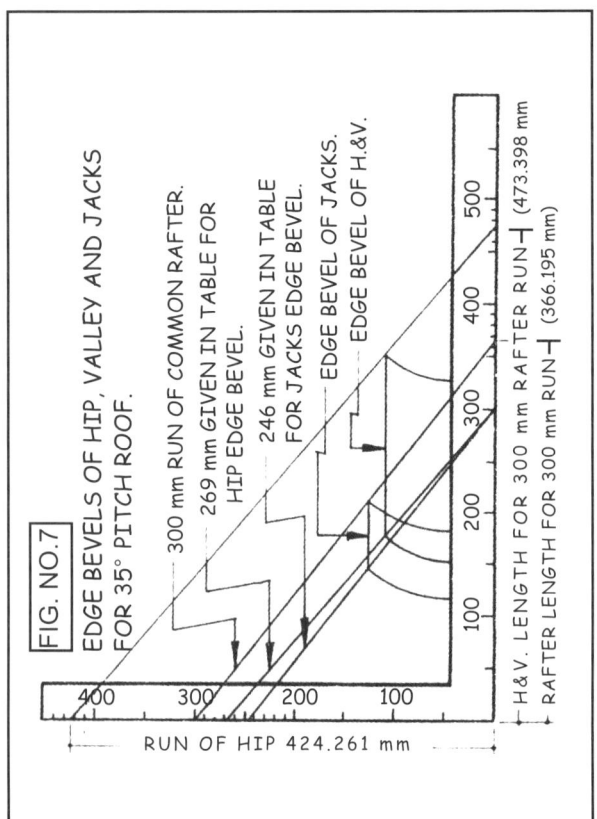

CHAPTER 4

Edge Bevels of Hips, Valleys and Jacks

To find the edge bevel of the hips and valleys. we use the full length of the hip on one arm of the square, and the full run of the hip on the other arm, and strike a diagonal line from these two points, see Fig. 9. The angle formed by that diagonal line and the arm of the square on which the length of the hip is used, is the angle of the edge bevel. For the edge bevel on the jack rafters, we use the full length of the common rafter on one arm of the square, and the full run of the common rafter on the other arm, and proceed in the same way as for the edge bevel of the hip.

But to use those figures, the full length, and full run, we first have to find them, as explained in Chapter 2. To do that with every roof when the total run, rise and length could be any odd measurement, would be very laborious, so we break it down and use 300 mm on the blade of the square, instead of the full lengths, and the run for 300 mm of its length, only, on the tongue of the square.

The same applies for the edge bevel of the jack rafters, using 300 mm on the blade of the square, and the run for 300 mm of its length on the tongue of the square, in each case marking on the blade for the edge bevel.

The same figures are used for the edge bevel of the purlin, as for the edge bevel of the jacks, but we mark along the tongue instead of the blade. In other words, the edge bevel of the purlin is square with the edge bevel of the jacks.

For the plumb or side bevel of the purlin, we use the total rise of the common rafter on the tongue of the square, and the total length of the common rafter on the blade, and mark along the tongue, or, to use the tables, 300 mm of the length of the common rafter on the blade of the square, and the rise for 300 mm of its length, on the tongue, marking along the tongue for cutting.

To find the backing bevel of the hip, seldom used now, use the total rise of the roof on the tongue of the square, and the total length of the hip on the blade, the angle formed by the diagonal of these two points and the tongue of the square, is the angle of the hip backing bevel. But to use the tables again, use 300 mm of the length of the hip on the blade of the square and the rise for 300 mm of its length, on the tongue of the square. It will not be possible to use these figures on the end of the hip timber, so mark the bevel on the side of the timber and transfer it to the end of the timber with an ordinary bevel.

All the figures on the tongue of the square to use in conjunction with 300 mm on the blade, to give all the foregoing bevels, have been calculated and compiled to form The Carpenter's Metric Roofing Ready Reckoner, thus eliminating time and labour involved in finding the various total lengths otherwise required. Furthermore, the total length of a hip or rafter would be such that it would not be possible to use their full lengths on the square, except at a considerably reduced scale.

The difference or diminish in the length of jack rafters is quite simple with this method of roofing, as the spacing from centre to centre represents the same amount of run of the common rafter. In other words 300 mm centres is the same as 300 mm run.

For the spacing of 450 mm centre to centre,

add the length for 400 mm and the length for 50 mm together, which for a 35° pitch is 488.26 plus 61.032 mm and the total of 549.292 mm equals the diminish or difference in length of jacks for a 35° pitch roof with a rafter spacing of 450 mm. 0.292 can be ignored and the diminish read as 549 mm, as given on the 35° table.

The difference in the length of jacks from 300 mm by 50 mm to 600 mm spacing is given on each table to eliminate even this simple mathematical sum.

With this method of roofing with the "Steel Square", the necessity of calculating the total lengths, total rise and total run of both rafters and hips, which could be any odd figure, and using those figures to obtain the various bevels as explained in this chapter, is dispensed with, as also is the need to memorise which figures to use with which, and on which arm of the square to mark for cutting. All this information is given in tabulated form in the ready reckoner tables at the end of this volume.

If the square is used according to the instructions, and the tables are used for calculating the lengths of rafters, hips and valleys etc., any carpenter should be able to cut all roof timbers before the building has even reached wall plate level, as any slight discrepancy in the width of the building could be adjusted when fixing the wall plates.

" . . . I had one fourteen years ago . . . it is the best roofing book I have ever held in my hand."
(G. M., Stoke-on-Trent, Staffs.)

"I purchased your Ready Reckoner in 1948 . . . I hope this valuable book is still obtainable."
(G. J ., Menai Bridge, Anglesey).

". . . I have just lost the one I had and I am lost without it. I must at this point thank you for a really wonderful little book, a book that is a boon to us carpenters. To be able to walk on to a site with confidence and put on a roof without any ado, is really the envy of a good many carpenters."
(S. P. C., Marske-by-the-Sea)"

"Your two books on roofing are the best I have come across.
(S. G ., Co. Down, N.I.

"I have never seen an easier system than yours."
(H. N. W., Reading, Berks)

"I have been using your Ready Reckoner now for the past five years and find that it is now an essential part of my tool kit."
(G. R. W., Trowbridge, Wilts.)

" . . . your Ready Reckoner is unbeatable. There is nothing to equal it available in New Zealand."
(D. D ., New Zealand)

"I have the fourth edition which you sent about eight years ago. It is a splendid book."
(E. B ., Hyde, Cheshire)

CHAPTER 5

Marking and Cutting the Roof Timbers

Having taken delivery of the timber - the necessary lengths of which can be "taken off" from the ready reckoner tables for the appropriate roof pitch - select a good straight length for the pattern rafter from which all the common rafters and jack rafters should be marked and cut.

The simplest way of marking the pattern rafter is illustrated in Fig. 10. First, mark the plumb level at one end of the timber; from the longest point of that mark measure along the back of the rafter the length obtained from the appropriate table, less the allowance for the thickness of the ridge (the full thickness for a lean-to roof and half the thickness for a span or coupled roof) and from that length point, mark the seat bevel.

In the case of rafters that extend beyond the wall plate to form an overhang or soffit, instead of marking the seat bevel from the point marking the length, as last described, mark another plumb bevel; now mark the seat

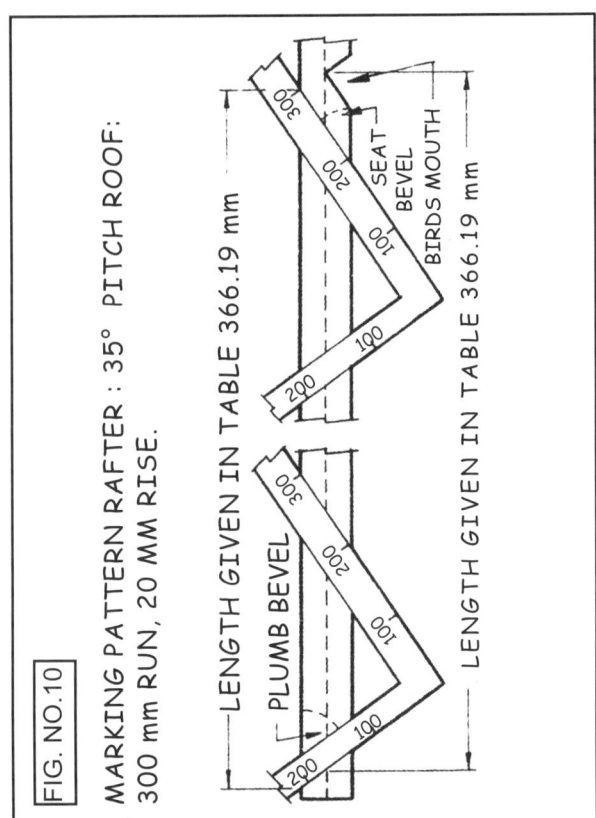

FIG. NO.10

MARKING PATTERN RAFTER : 35° PITCH ROOF: 300 mm RUN, 20 MM RISE.

bevel, intersecting the plumb bevel at the required depth of the hakes or bird's mouth; square the lines over the edge, and the rafter is ready for cutting. It is better to mark one rafter as a pattern, checking to make sure it is correct, then mark all the other rafters from that pattern.

The foot of the rafters can be cut to the plumb and seat bevel at this stage, if so required, but I prefer to leave the overhanging portion long enough to be cut off to the required overhang after all the rafters and hips are fixed, thus ensuring a straight line for fixing the facia boards.

Before marking the hips it must be decided which method of construction for the top joint is to be used; there are at least three alternatives, different districts favouring one or the other.

Fig. 11 illustrates the three methods A, B and C. With method A, which I prefer, no deduction for the ridge is necessary and the centres of the ridge, rafters and hips all converge at one point, as they should.

The central end jack rafter, between the two hips, is a full length rafter, less half the thick-

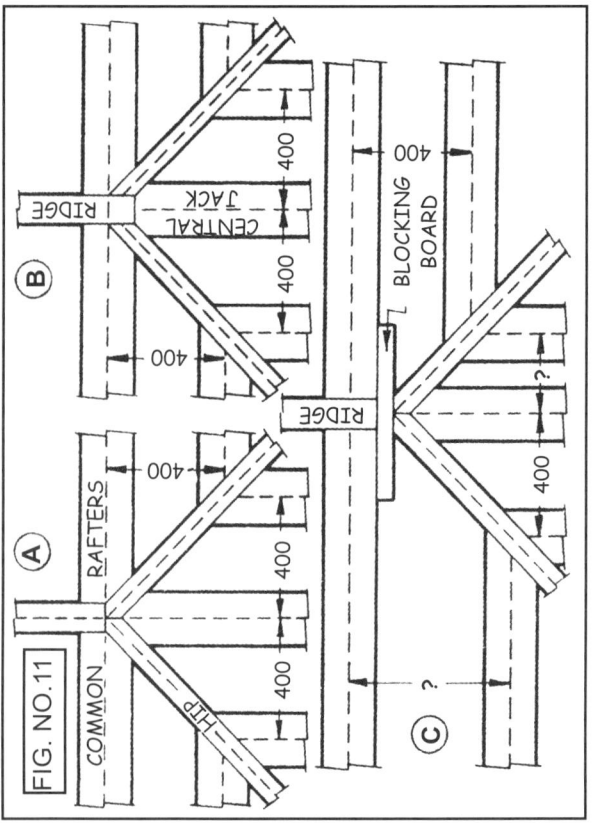

ness of the hip. The diminish for the subsequent jacks, as given in the appropriate table, is measured from the shortest point of the edge bevel of the central jack, to the shortest point of the edge bevel of the next jack.

All jack rafters will then come opposite each other and at correct centres on each hip.

With method 8, allowance for half the thickness of the ridge, must be deducted from the length given in the appropriate table. Method C is sometimes used, but it does rather complicate the positioning of the first jack rafters. The spacing between the first common rafter and the first jack rafter, is greater than the subsequent spaces, because the first common rafters have been moved away from the central point by half their thickness, plus the thickness of the planted-on blocking board, as illustration C (Fig. 11) shows.

As previously stated, I prefer method A. Fig. 12 illustrates the marking of the hip timber for this method. Having decided the back or top edge of the hip (the top edge of all roof timbers should be the hollow edge) mark on that edge the edge bevel and from the edge bevel mark the plumb bevel on both

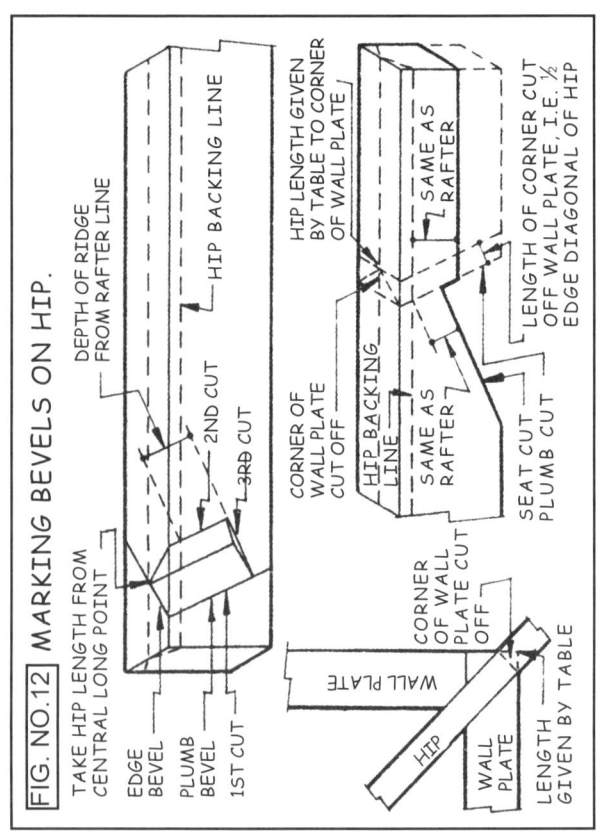

sides of the timber, and cut. Now, from the edge bevel, i.e. the centre of the hip, mark the edge bevel in the opposite direction and mark the plumb bevel on the other side to the required depth to take the ridge. That depth is the distance from the rafter line on the ridge to the underside of the ridge timber and, on the hip, is measured from the hip backing line, or rafter line, on the side of the hip. The hip backing line should always be marked on both sides of the hip timber, even if the hip is not going to be bevelled. Now from that depth line, draw a line square with the plumb bevel, then the second and third cuts can be made, completing the top cuts of the hip.

From the longest point of the completed top cut, i.e. centre of the edge of the hip, measure the length as given in the appropriate table (no deduction for the ridge is necessary) and square a line across; now, measure back from that line the amount of the corner to be cut off the wall plate (see Fig. 12) and square another line across; from that second line, mark the plumb bevel on both sides of the hip, then mark the seat bevel intersecting the plumb bevel at the required depth of the hakes or bird's mouth.

In practice, it is wise to check the length of each hip with rod or tape on the actual building after the ridge and first rafters are in position, before marking and cutting the bottom bevels, in case the building is not quite square ; any slight difference in the lengths would not appreciably affect the bevels.

"I have been using your Roofing Ready Reckoner for several years now and find it absolutely invaluable."
(R . G., Oxford)

", . . I must add the book is wonderful and so easy to understand, by far the best method for roofing I have come across."
(R. E. W ., Christchurch, Hants)

" . . . having used one for the past ten years , it is the finest book that I have ever seen and is a real asset to any carpenter."
(J. A. D., Breaston, Derbyshire)

"I think it is great, it 's the best book I have ever read. "
(B. C., Omagh, N.1.)

" . . . thank you for your unbeatable Carpenter 's Roofing Ready Reckoner."
(F. F., New Zealand)

" . . . I think it is a marvellous little book."
(W. S., Dublin, Eire)

". . . What a boon I found . . . the Carpenter·s Roofing Ready Reckoner which I bought in 1950, it has been a wonderful time saver."
(E. H. R., Romford, Essex)

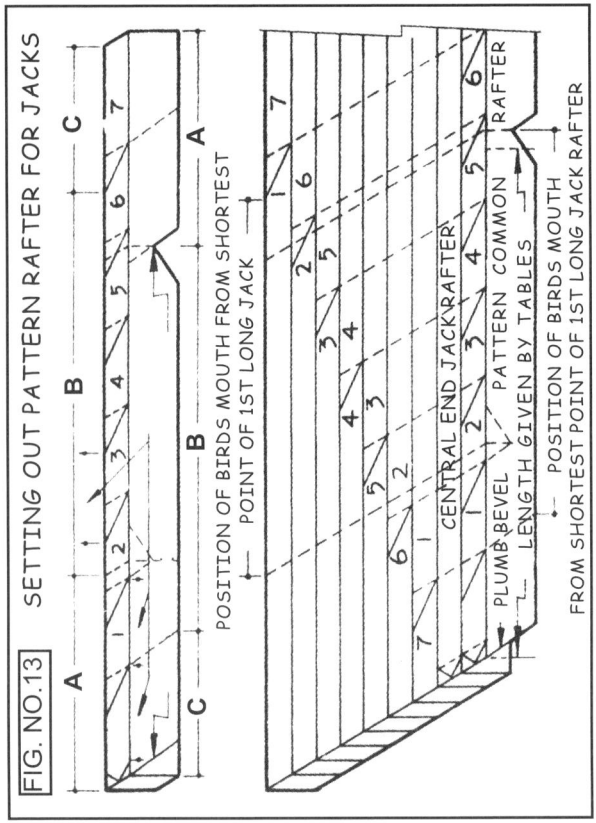

CHAPTER 6

Marking and Cutting Jack Rafters and Purlins

Many carpenters experience difficulty in obtaining the correct length of the first long jack rafter, they measure the diminish from the long point of the pattern common rafter, to the longest point of the edge bevel for the first long jack rafter, instead of to the centre of the edge bevel. This results in the first long jack rafter being short and the spacing from the first common rafter being increased.

Fig. 13 illustrates the correct way to mark the pattern common rafter for diminish of jacks, and how to transfer those marks to the timber for each successive jack rafter, so that one cut makes two jack rafters. If a hip has, say, seven jack rafters each side, then, seven lengths of timber the same length as the pattern rafter, should cut fourteen jacks, i.e. a complete set for each hip.

Remember, the diminish for the first long jack rafter is measured from the long point of the common rafter, to the centre of the edge

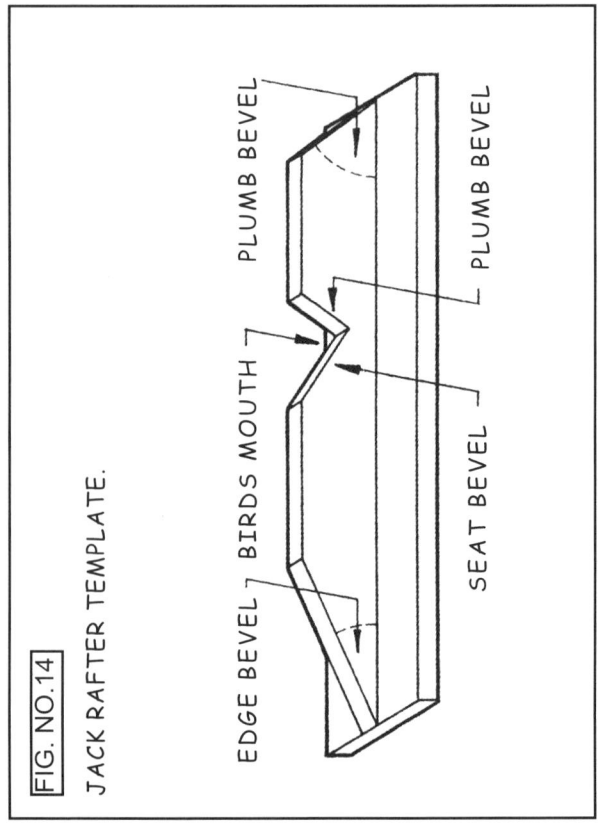

FIG. NO.14
JACK RAFTER TEMPLATE.

bevel, all subsequent jacks can be measured for diminish from long point to long point; or short point to short point; but the first long jack diminish is always measured from the longest point of the common rafter to the centre of the edge bevel.

It should never be necessary to take the square up on to the roof, all marking and cutting should be done on the ground. Use the square to mark the pattern common rafter, each hip and valley, and the purlins. For the jack rafters, make a template about 300 mm long, and slightly wider than the timber being used for the rafters; cut the rafter plumb bevel at one end, and the jack's edge bevel at the other, and the bird's mouth in the centre, using the square to obtain these bevels. Now screw a fence on to the back edge of this double bevel template as illustrated in Fig 14.

Fig.15 illustrates the purlin bevels; on the edge at one end of the timber, mark the edge bevel and the plumb bevel, and, if the depth of the purlin timber is such that the bottom edge comes below the bottom edge of the hip, then at the required depth, the purlin should continue, to mitre at the centre of the hip as illustrated in the inset in Fig. 15.

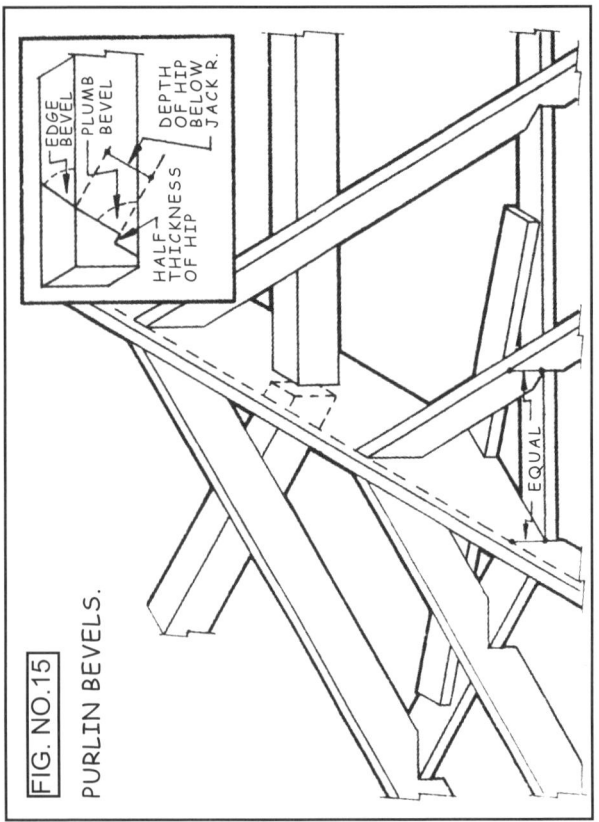

FIG. NO.15
PURLIN BEVELS.

I always like to keep the top edge of the purlin slightly lower than the bottom edge of the rafters, as much as 12.5 mm, then when the rafters are nailed down to the purlin, they are slightly and uniformly hollow. This gives a "springing" for the slates or tiles. and they lie better; nothing looks worse than a roof with a hump in it.

———

" . . . I have read many of the testimonials you have received and I fully endorse what they say, they are wonderful books."
(E. H. R ., Romford. Essex)

" . . . I bought one from you over ten years ago . . . and was thoroughly satisfied with it."
(P. H., St. Mary Cray, Kent)

". . . it is so simple and has such a direct approach . . . saving an immense amount of time."
(G . P ., Chesterfield, Derbyshire)

"...I've had one of these for the past twelve years ... it's the best I've had."
(E. A. C ., Camberley. Surrey)

" . . . I have been using your book for approximately 15 years . . . I have found same most helpful."
(H . G. G ., Combe Martin, Devon)

CHAPTER 7

Assembling and Erecting the Roof Timbers

Now that all the common rafters and jack rafters have been cut, and the hips top cut only, the wall plates should be fixed. Make sure that they are in their correct position, level, parallel, and square, making at this stage any adjustment necessary to attain this. From the outside edge of the end wall plates mark along the main wall plates half the total width of the building, from wall plate to wall plate; that mark is the centre of the first common rafter. From that first common rafter, at each end of the building, set out the position of the remainder of the common rafters and jack rafters.

The centre of the end wall plates is the centre of the central jack rafter, (see Fig. 11A), and from that central end jack rafter and towards each hip, the position of the remaining jack rafters should be set out on the wall plates. The ceiling joist should now be fixed to tie in the wall plates, otherwise the thrust of the rafters would force them off the walls. If the ceiling joist is to be fixed higher than the wall plates and nailed to the side of the rafters, then the wall plates should be temporarily tied, to secure them in position until all the rafters are fixed and the ceiling joist can be nailed to them.

The ridge should now be cut to the exact length, i.e. from the centre to the centre of the first common rafters, and the position of the rafters should be set out to correspond with the setting out on the main wall plates, working from each end towards the centre. The rafter line should be marked on each side of the ridge, 37 mm to 44 mm from the top edge, according to the thickness of the slating or tiling battens.

The first pair of common rafters (for each end of the building) which have been coupled together at the correct width of the building, i.e. out to out of the wall plates, with the tops of the rafters held apart for the thickness of the ridge, with a batten fixed at the correct depth to support the ridge, should now be erected and fixed in their correct position. One or two more intermediate pairs of rafters similarly coupled, should also be erected to support the ridge, which can now be placed in position. Make sure that the rafters are fixed in the position corresponding with the setting out on the wall plates and ridge, and temporarily braced so that the distance from the end of the ridge to the outside edge of the end wall plate is the same at each end of the building (this should be checked with a long batten or tape); the rafters will then automatically be plumb.

The length of the hips should now be checked with a tape or long batten; if the building is square, they should all be the same length, as given in the tables in the ready reckoner. There may be from 12 to 14 mm difference if the building is not square, but this would not appreciably affect the bevels. Each hip should now be marked for length as measured, and the bottom bevels marked and cut as explained in Fig. 12 and they will then be ready for fixing in position.

Next fix the central end jack rafters, these central end jacks are the same length as the common rafter, less the half-thickness of the hip measured on the diagonal, both these central end jack rafters should be the same length. The purlins should next be fixed and then the remaining common rafters and jack rafters, followed by the facia and soffit. The roof is now ready for battening.

" . . . it is quite the best book on roofing I have ever had."

(D . C ., Eastbourne, Sussex)

". . . It is the best and simplest book of its kind that I have found."

(J . F. S ., Wooton. I.0.W.)

CHAPTER 8

Battening Roof for Slates

In some districts the slaters or tilers always batten the roof to suit themselves, and they can generally be safely left to do so; but when the carpenter is called on to do it, as he often is in the country districts, it is just as well to know the correct and easiest way to do it.

Assuming the roof is to be covered with 250 mm x 500 mm slates with a 100 mm bond (see Fig. 16) the centre of the first batten would be 500 mm (less the office or overhang) from the outside edge of the facia board. This first batten must be fixed straight and true, as all the other battens are governed by this one, gauging from top to top of each subsequent batten.

To find the correct spacing for the battens, deduct the bond from the length of the slate and divide by two, e.g. 500 mm less 100 mm is 400 mm, which, divided by two is 200 mm, and that is the spacing from top to top of each batten. The last batten should be fixed tight against the ridge, and from the top of that last

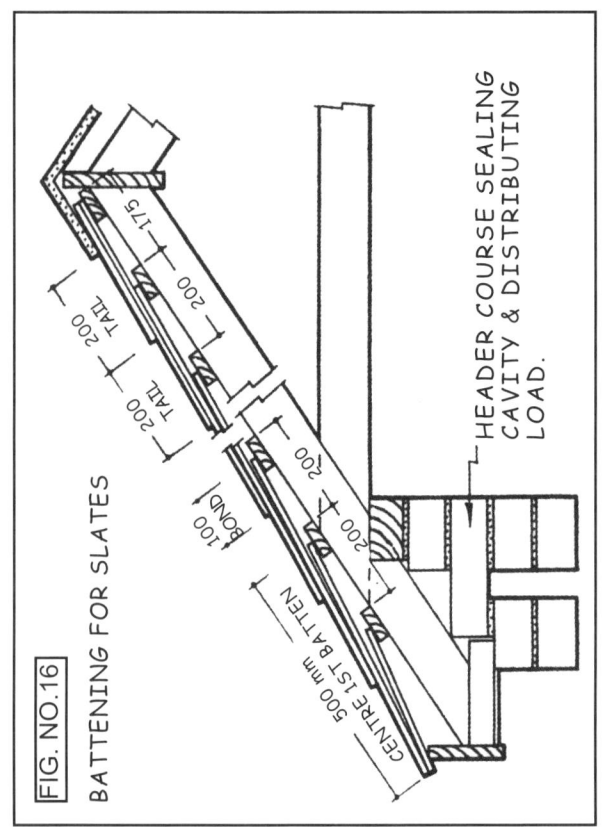

batten to the top of the batten immediately below, should be 25 mm less than the other battens, in this case 175 mm instead of 200 mm.

This is important, as it ensures the same amount of tail of slate showing below the ridge tiles as the remainder of the roof, instead of that narrow little bit of tail one so often sees, which looks so bad.

Some carpenters use a wooden thumb gauge, working from top to top of the battens, but with this method it is so easy to lose or gain a fraction each time, entailing the necessity of constantly sighting and checking the battens for being straight.

I prefer to cut a batten the exact length from the ridge to the top of the first batten fixed, and on that batten, mark the position of the remaining battens. Having cut the batten, square a line across 175 mm from the top end, i.e. the position of the top of the last batten prior to the batten fixed tight against the ridge. Now take a pair of dividers, set them at 200 mm and 'walk' them along the batten starting from the bottom end; the last point must finish on the 175 mm mark at the top of the batten. To attain that, it may be necessary to either gain or lose a fraction each time; this can only be ascertained by trial and error, adjusting the dividers as necessary.

Having found the correct spacing to show 75 mm for the last batten below the ridge, which has already been squared across, square lines to mark the top of each batten. Now place this batten edgeways on the back of the first common rafter, resting the bottom end of the batten on top of the first batten already fixed, and transfer the lines marking the top of each batten, to the first common rafter at each end of the building; do the same on one or two intermediate rafters if necessary. Now, take a long line (dipped in a wet solution coloured with raddle or lamp black) draw it tight on the batten marks on the first common rafters, then flick it at the centre and all the rafters will be correctly marked with the position of the top edge of all the battens. This method is far quicker, easier, and more accurate than using a thumb gauge, which one is for ever dropping, particularly if the boss happens to be anywhere near; the need for continually sighting the battens for being straight, is eliminated.

The battens should mitre on the centre of the hips, whether the hip is back bevelled or not, if the hip is not back bevelled it throws the battens and thus the slates, a little high on the hip, which is an advantage inasmuch that rainwater is conducted away from the hip joint of the slates.

If the roof is to be tiled, the spacing of the battens will vary according to the type of tiles used, but the principle and the method of battening the roof is the same. As there are so many different patterns and types of tiles on the market it is not possible to give details of the spacing of the battens here; these details should be ascertained from the makers.

It is not possible nor indeed advisable to try to foresee and describe every contingency likely to be met with in roofing otherwise these articles would be far too long and confusing; only the straightforward hipped roof has been dealt with.

The method of approach will vary in different districts and with different carpenters but, the principle remains the same, and as the younger carpenter gains knowledge and experience, the solving of problems and difficulties will become much easier and more interesting, and he will learn to apply these principles to the more complicated roofs with equal success and greater satisfaction.

The following tables can be used with any type of roofing square which has millimetre graduations on both blade and tongue.

All the testimonials quoted, are from the many hundreds received from users of The *Carpenter's Roofing Ready Reckoner* published so long ago; as the same method is used in The Carpenter's Metric Roofing Ready Reckoner (but with additional information) I hope and believe that this new, revised edition will prove equally as popular and even more helpful.

METRIC ROOFING READY RECKONER

FOR EVERY 2 ½° PITCH

FROM 17 ½° TO 60°

$17\frac{1}{2}° = 94.5$ mm RISE IN 300 mm RUN

RUN OF R	LENGTH OF RAFTER	LENGTH OF H&V	ROOF MEMBER AND BEVEL		FIGURES USED ON BLADE TONGUE		MARK CUT ON
1	1·0484	1·4488	R	PLUMB	300	94·5	TONGUE
2	2·0968	2·8976	R	SEAT	300	94·5	BLADE
3	3·1453	4·3464	H&V	PLUMB	424	94·5	TONGUE
4	4·1937	5·7952	H&V	SEAT	424	94·5	BLADE
5	5·2421	7·2439	H&V	EDGE	300	293·0	BLADE
6	6·2905	8·6927	HIP	BACK'G	300	65·0	TONGUE
7	7·3389	10·141	PURLIN	PLUMB	300	90·0	TONGUE
8	8·3874	11·59	PURLIN	EDGE	300	286·0	TONGUE
9	9·4358	13·039	JACKS	EDGE	300	286·0	BLADE
10	10·484	14·488					

MOVE DECIMAL POINT 1, 2, OR 3 PLACES TO RIGHT AND UNITS BECOME TENS, HUNDREDS OR THOUSANDS 0·4 & UNDER IGNORE. 0·5 & OVER ADD 1 TO UNITS.

DIFFERENCE IN JACKS
300 mm CENTRES 314·5 mm
350 mm CENTRES 367 mm
400 mm CENTRES 419·5 mm
450 mm CENTRES 472 mm
500 mm CENTRES 524 mm
550 mm CENTRES 576·5 mm
600 mm CENTRES 629 mm

20° = 109·5 mm RISE IN 300 mm RUN

RUN OF R	LENGTH OF RAFTER	LENGTH OF H&V	ROOF MEMBER AND BEVEL		FIGURES USED ON BLADE TONGUE		MARK CUT ON
1	1·0645	1·4605	R	PLUMB	300	109·5	TONGUE
2	2·129	2·9211	R	SEAT	300	109·5	BLADE
3	3·1935	4·3816	H&V	PLUMB	424	109·5	TONGUE
4	4·258	5·8422	H&V	SEAT	424	109·5	BLADE
5	5·3225	7·3027	H&V	EDGE	300	291·0	BLADE
6	6·3871	8·7632	HIP	BACK'G	300	75·0	TONGUE
7	7·4516	10·224	PURLIN	PLUMB	300	103·0	TONGUE
8	8·5161	11·684	PURLIN	EDGE	300	281·0	TONGUE
9	9·5806	13·145	JACKS	EDGE	300	281·0	BLADE
10	10·645	14·605					

MOVE DECIMAL POINT 1, 2, OR 3 PLACES TO RIGHT AND UNITS BECOME TENS, HUNDREDS OR THOUSANDS 0·4 & UNDER IGNORE. 0·5 & OVER ADD 1 TO UNITS.

DIFFERENCE IN JACKS
300 mm CENTRES 319·5 mm
350 mm CENTRES 372·5 mm
400 mm CENTRES 425·5 mm
450 mm CENTRES 479 mm
500 mm CENTRES 532 mm
550 mm CENTRES 585·5 mm
600 mm CENTRES 638·5 mm

22½° = 124 mm RISE IN 300 mm RUN

RUN OF R	LENGTH OF RAFTER	LENGTH OF H&V	ROOF MEMBER AND BEVEL		FIGURES USED ON BLADE TONGUE		MARK CUT ON
1	1·0823	1·4739	R	PLUMB	300	124·0	TONGUE
2	2·1647	2·9479	R	SEAT	300	124·0	BLADE
3	3·2471	4·4218	H&V	PLUMB	424	124·0	TONGUE
4	4·3294	5·8957	H&V	SEAT	424	124·0	BLADE
5	5·4117	7·3696	H&V	EDGE	300	288·0	BLADE
6	6·4941	8·8436	HIP	BACK'G	300	84·0	TONGUE
7	7·5764	10·317	PURLIN	PLUMB	300	115·0	TONGUE
8	8·6588	11·791	PURLIN	EDGE	300	277·0	TONGUE
9	9·7411	13·265	JACKS	EDGE	300	277·0	BLADE
10	10·823	14·739					

MOVE DECIMAL POINT 1, 2, OR 3 PLACES TO RIGHT AND UNITS BECOME TENS, HUNDREDS OR THOUSANDS 0·4 & UNDER IGNORE. 0·5 & OVER ADD 1 TO UNITS.

DIFFERENCE IN JACKS
300 mm CENTRES 324·5 mm
350 mm CENTRES 378·5 mm
400 mm CENTRES 432·5 mm
450 mm CENTRES 487 mm
500 mm CENTRES 541 mm
550 mm CENTRES 595 mm
600 mm CENTRES 649 mm

25° = 140 mm RISE IN 300 mm RUN

RUN OF R	LENGTH OF RAFTER	LENGTH OF H&V	ROOF MEMBER AND BEVEL		FIGURES USED ON BLADE TONGUE		MARK CUT ON
1	1·1035	1·4888	R	PLUMB	300	140·0	TONGUE
2	2·2069	2·9777	R	SEAT	300	140·0	BLADE
3	3·3104	4·4665	H&V	PLUMB	424	140·0	TONGUE
4	4·4139	5·9554	H&V	SEAT	424	140·0	BLADE
5	5·5174	7·4442	H&V	EDGE	300	285·0	BLADE
6	6·6209	8·9331	HIP	BACK'G	300	94·0	TONGUE
7	7·7244	10·422	PURLIN	PLUMB	300	127·0	TONGUE
8	8·8278	11·911	PURLIN	EDGE	300	272·0	TONGUE
9	9·9313	13·399	JACKS	EDGE	300	272·0	BLADE
10	11·035	14·888					

MOVE DECIMAL POINT 1, 2, OR 3 PLACES TO RIGHT AND UNITS BECOME TENS, HUNDREDS OR THOUSANDS 0·4 & UNDER IGNORE. 0·5 & OVER ADD 1 TO UNITS.

DIFFERENCE IN JACKS
300 mm CENTRES 331 mm
350 mm CENTRES 386 mm
400 mm CENTRES 441 mm
450 mm CENTRES 496·5 mm
500 mm CENTRES 552 mm
550 mm CENTRES 607 mm
600 mm CENTRES 662 mm

$27\frac{1}{2}° = 156$ mm RISE IN 300 mm RUN

RUN OF R	LENGTH OF RAFTER	LENGTH OF H&V	ROOF MEMBER AND BEVEL		FIGURES USED ON BLADE TONGUE		MARK CUT ON
1	1·1271	1·5068	R	PLUMB	300	156·0	TONGUE
2	2·2542	3·0135	R	SEAT	300	156·0	BLADE
3	3·3813	4·5203	H&V	PLUMB	424	156·0	TONGUE
4	4·5084	6·0271	H&V	SEAT	424	156·0	BLADE
5	5·6355	7·5339	H&V	EDGE	300	281·0	BLADE
6	6·7627	9·0407	HIP	BACK'G	300	104·0	TONGUE
7	7·8898	10·547	PURLIN	PLUMB	300	139·0	TONGUE
8	9·0169	12·054	PURLIN	EDGE	300	266·0	TONGUE
9	10·144	13·561	JACKS	EDGE	300	266·0	BLADE
10	11·271	15·068					

MOVE DECIMAL POINT 1, 2, OR 3 PLACES TO RIGHT AND UNITS BECOME TENS, HUNDREDS OR THOUSANDS 0·4 & UNDER IGNORE. 0·5 & OVER ADD 1 TO UNITS.

DIFFERENCE IN JACKS
300 mm CENTRES 338 mm
350 mm CENTRES 394·5 mm
400 mm CENTRES 451 mm
450 mm CENTRES 507 mm
500 mm CENTRES 563·5 mm
550 mm CENTRES 620 mm
600 mm CENTRES 676 mm

30° = 173 mm RISE IN 300 mm RUN

RUN OF R	LENGTH OF RAFTER	LENGTH OF H&V	ROOF MEMBER AND BEVEL		FIGURES USED ON BLADE TONGUE		MARK CUT ON
1	1·1543	1·5274	R	PLUMB	300	173·0	TONGUE
2	2·3087	3·055	R	SEAT	300	173·0	BLADE
3	3·462	4·5823	H&V	PLUMB	424	173·0	TONGUE
4	4·6173	6·1098	H&V	SEAT	424	173·0	BLADE
5	5·7716	7·6372	H&V	EDGE	300	278·0	BLADE
6	6·926	9·1647	HIP	BACK'G	300	11·4	TONGUE
7	8·0803	10·692	PURLIN	PLUMB	300	150·0	TONGUE
8	9·2346	12·219	PURLIN	EDGE	300	260·0	TONGUE
9	10·389	13·748	JACKS	EDGE	300	260·0	BLADE
10	11·543	15·274					

MOVE DECIMAL POINT 1, 2, OR 3 PLACES TO RIGHT AND UNITS BECOME TENS, HUNDREDS OR THOUSANDS 0·4 & UNDER IGNORE. 0·5 & OVER ADD 1 TO UNITS.

DIFFERENCE IN JACKS
300 mm CENTRES 346 mm
350 mm CENTRES 404 mm
400 mm CENTRES 461·5 mm
450 mm CENTRES 519 mm
500 mm CENTRES 577 mm
550 mm CENTRES 634·5 mm
600 mm CENTRES 692·5 mm

$32\frac{1}{2}° = 191$ mm RISE IN 300 mm RUN

RUN OF R	LENGTH OF RAFTER	LENGTH OF H&V	ROOF MEMBER AND BEVEL		FIGURES USED ON BLADE TONGUE		MARK CUT ON
1	1·1851	1·5512	R	PLUMB	300	191·0	TONGUE
2	2·3703	3·1025	R	SEAT	300	191·0	BLADE
3	3·5554	4·6537	H&V	PLUMB	424	191·0	TONGUE
4	4·7406	6·205	H&V	SEAT	424	191·0	BLADE
5	5·9257	7·7562	H&V	EDGE	300	273·0	BLADE
6	7·1109	9·3075	HIP	BACK'G	300	124·0	TONGUE
7	8·2961	10·859	PURLIN	PLUMB	300	161·0	TONGUE
8	9·4812	12·41	PURLIN	EDGE	300	253·0	TONGUE
9	10·666	13·961	JACKS	EDGE	300	253·0	BLADE
10	11·851	15·512					

MOVE DECIMAL POINT 1, 2, OR 3 PLACES TO RIGHT AND UNITS BECOME TENS, HUNDREDS OR THOUSANDS 0·4 & UNDER IGNORE. 0·5 & OVER ADD 1 TO UNITS.

DIFFERENCE IN JACKS
300 mm CENTRES 355·5 mm
350 mm CENTRES 414·5 mm
400 mm CENTRES 474 mm
450 mm CENTRES 533 mm
500 mm CENTRES 592·5 mm
550 mm CENTRES 651·5 mm
600 mm CENTRES 711 mm

35° = 210 mm RISE IN 300 mm RUN

RUN OF R	LENGTH OF RAFTER	LENGTH OF H&V	ROOF MEMBER AND BEVEL		FIGURES USED ON BLADE TONGUE		MARK CUT ON
1	1·2206	1·5779	R	PLUMB	300	210·0	TONGUE
2	2·4413	3·1559	R	SEAT	300	210·0	BLADE
3	3·6619	4·7339	H&V	PLUMB	424	210·0	TONGUE
4	4·8826	6·3118	H&V	SEAT	424	210·0	BLADE
5	6·1032	7·8898	H&V	EDGE	300	269·0	BLADE
6	7·3239	9·4678	HIP	BACK'G	300	134·0	TONGUE
7	8·5445	11·046	PURLIN	PLUMB	300	172·0	TONGUE
8	9·7652	12·624	PURLIN	EDGE	300	245·0	TONGUE
9	10·986	14·202	JACKS	EDGE	300	245·0	BLADE
10	12·206	15·779					

MOVE DECIMAL POINT 1, 2, OR 3 PLACES TO RIGHT AND UNITS BECOME TENS, HUNDREDS OR THOUSANDS 0·4 & UNDER IGNORE. 0·5 & OVER ADD 1 TO UNITS.

DIFFERENCE IN JACKS
300 mm CENTRES 366 mm
350 mm CENTRES 427 mm
400 mm CENTRES 488 mm
450 mm CENTRES 549 mm
500 mm CENTRES 610 mm
550 mm CENTRES 671 mm
600 mm CENTRES 732 mm

$37\frac{1}{2}° = 230·5$ mm RISE IN 300 mm RUN

RUN OF R	LENGTH OF RAFTER	LENGTH OF H&V	ROOF MEMBER AND BEVEL		FIGURES USED ON BLADE TONGUE		MARK CUT ON
1	1·2608	1·6094	R	PLUMB	300	230·5	TONGUE
2	2·5215	3·2189	R	SEAT	300	230·5	BLADE
3	3·7823	4·8283	H&V	PLUMB	424	230·5	TONGUE
4	5·043	6·4377	H&V	SEAT	424	230·5	BLADE
5	6·3038	8·0471	H&V	EDGE	300	264·0	BLADE
6	7·5646	9·6566	HIP	BACK'G	300	144·0	TONGUE
7	8·8253	11·266	PURLIN	PLUMB	300	182·0	TONGUE
8	10·086	12·875	PURLIN	EDGE	300	238·0	TONGUE
9	11·347	14·485	JACKS	EDGE	300	238·0	BLADE
10	12·608	16·094					

MOVE DECIMAL POINT 1, 2, OR 3 PLACES TO RIGHT AND UNITS BECOME TENS, HUNDREDS OR THOUSANDS 0·4 & UNDER IGNORE. 0·5 & OVER ADD 1 TO UNITS.

DIFFERENCE IN JACKS
300 mm CENTRES 378 mm
350 mm CENTRES 441 mm
400 mm CENTRES 504 mm
450 mm CENTRES 567 mm
500 mm CENTRES 630 mm
550 mm CENTRES 693 mm
600 mm CENTRES 756 mm

40° = 252 mm RISE IN 300 mm RUN

RUN OF R	LENGTH OF RAFTER	LENGTH OF H&V	ROOF MEMBER AND BEVEL		FIGURES USED ON BLADE TONGUE		MARK CUT ON
1	1·3059	1·6479	R	PLUMB	300	252·0	TONGUE
2	2·6119	3·2958	R	SEAT	300	252·0	BLADE
3	3·9179	4·9437	H&V	PLUMB	424	252·0	TONGUE
4	5·2239	6·5916	H&V	SEAT	424	252·0	BLADE
5	6·5298	8·2395	H&V	EDGE	300	258·0	BLADE
6	7·8358	9·8874	HIP	BACK'G	300	154·0	TONGUE
7	9·1418	11·535	PURLIN	PLUMB	300	193·0	TONGUE
8	10·448	13·183	PURLIN	EDGE	300	229·0	TONGUE
9	11·754	14·831	JACKS	EDGE	300	229·0	BLADE
10	13·059	16·479					

MOVE DECIMAL POINT 1, 2, OR 3 PLACES TO RIGHT AND UNITS BECOME TENS, HUNDREDS OR THOUSANDS 0·4 & UNDER IGNORE. 0·5 & OVER ADD 1 TO UNITS.

DIFFERENCE IN JACKS
300 mm CENTRES 392 mm
350 mm CENTRES 457 mm
400 mm CENTRES 522 mm
450 mm CENTRES 587·5 mm
500 mm CENTRES 653 mm
550 mm CENTRES 718 mm
600 mm CENTRES 783·5 mm

$42\frac{1}{2}° = 275$ mm RISE IN 300 mm RUN

RUN OF R	LENGTH OF RAFTER	LENGTH OF H&V	ROOF MEMBER AND BEVEL		FIGURES USED ON BLADE TONGUE		MARK CUT ON
1	1·3565	1·6853	R	PLUMB	300	275·0	TONGUE
2	2·7131	3·3706	R	SEAT	300	275·0	BLADE
3	4·0696	5·0558	H&V	PLUMB	424	275·0	TONGUE
4	5·4261	6·7411	H&V	SEAT	424	275·0	BLADE
5	6·7826	8·4264	H&V	EDGE	300	252·0	BLADE
6	8·1392	10·112	HIP	BACK'G	300	164·0	TONGUE
7	9·4957	11·797	PURLIN	PLUMB	300	203·0	TONGUE
8	10·852	13·482	PURLIN	EDGE	300	221·0	TONGUE
9	12·209	15·167	JACKS	EDGE	300	221·0	BLADE
10	13·565	16·853					

MOVE DECIMAL POINT 1, 2, OR 3 PLACES TO RIGHT AND UNITS BECOME TENS, HUNDREDS OR THOUSANDS 0·4 & UNDER IGNORE. 0·5 & OVER ADD 1 TO UNITS.

DIFFERENCE IN JACKS
300 mm CENTRES 407 mm
350 mm CENTRES 475 mm
400 mm CENTRES 542·5 mm
450 mm CENTRES 610·5 mm
500 mm CENTRES 678 mm
550 mm CENTRES 836 mm
600 mm CENTRES 814 mm

45° = 300 mm RISE IN 300 mm RUN

RUN OF R	LENGTH OF RAFTER	LENGTH OF H&V	ROOF MEMBER AND BEVEL		FIGURES USED ON BLADE TONGUE		MARK CUT ON
1	1·4142	1·732	R	PLUMB	300	300·0	TONGUE
2	2·8284	3·4641	R	SEAT	300	300·0	BLADE
3	4·2426	5·1961	H&V	PLUMB	424	300·0	TONGUE
4	5·6568	6·9282	H&V	SEAT	424	300·0	BLADE
5	7·071	8·6602	H&V	EDGE	300	245·0	BLADE
6	8·4852	10·392	HIP	BACK'G	300	174·0	TONGUE
7	9·8994	12·124	PURLIN	PLUMB	300	212·0	TONGUE
8	11·314	13·856	PURLIN	EDGE	300	212·0	TONGUE
9	12·728	15·588	JACKS	EDGE	300	212·0	BLADE
10	14·142	17·32					

MOVE DECIMAL POINT 1, 2, OR 3 PLACES TO RIGHT AND UNITS BECOME TENS, HUNDREDS OR THOUSANDS 0·4 & UNDER IGNORE. 0·5 & OVER ADD 1 TO UNITS.

DIFFERENCE IN JACKS
300 mm CENTRES 424 mm
350 mm CENTRES 495 mm
400 mm CENTRES 565·5 mm
450 mm CENTRES 636 mm
500 mm CENTRES 707 mm
550 mm CENTRES 778 mm
600 mm CENTRES 848·5 mm

| \multicolumn{6}{|c|}{$47\frac{1}{2}° = 327{\cdot}75$ mm RISE IN 300 mm RUN} |

RUN OF R	LENGTH OF RAFTER	LENGTH OF H&V	ROOF MEMBER AND BEVEL		FIGURES USED ON BLADE TONGUE		MARK CUT ON
1	1·481	1·787	R	PLUMB	300	328·0	TONGUE
2	2·9621	3·5741	R	SEAT	300	328·0	BLADE
3	4·4431	5·3611	H&V	PLUMB	424	328·0	TONGUE
4	5·9242	7·1481	H&V	SEAT	424	328·0	BLADE
5	7·4052	8·9351	H&V	EDGE	300	237·0	BLADE
6	8·8862	10·722	HIP	BACK'G	300	184·0	TONGUE
7	10·367	12·509	PURLIN	PLUMB	300	221·0	TONGUE
8	11·848	14·296	PURLIN	EDGE	300	202·0	TONGUE
9	13·329	16·083	JACKS	EDGE	300	202·0	BLADE
10	14·81	17·87					

MOVE DECIMAL POINT 1, 2, OR 3 PLACES TO RIGHT AND UNITS BECOME TENS, HUNDREDS OR THOUSANDS 0·4 & UNDER IGNORE. 0·5 & OVER ADD 1 TO UNITS.

DIFFERENCE IN JACKS
300 mm CENTRES 444 mm
350 mm CENTRES 518 mm
400 mm CENTRES 592 mm
450 mm CENTRES 666·5 mm
500 mm CENTRES 740·5 mm
550 mm CENTRES 814·5 mm
600 mm CENTRES 888·5 mm

50° = 357·5 mm RISE IN 300 mm RUN

RUN OF R	LENGTH OF RAFTER	LENGTH OF H&V	ROOF MEMBER AND BEVEL		FIGURES USED ON BLADE TONGUE		MARK CUT ON
1	1·5554	1·8495	R	PLUMB	300	357·5	TONGUE
2	3·1108	3·6991	R	SEAT	300	357·5	BLADE
3	4·6662	5·5486	H&V	PLUMB	424	357·5	TONGUE
4	6·2216	7·3982	H&V	SEAT	424	357·5	BLADE
5	7·7769	9·2477	H&V	EDGE	300	230·0	BLADE
6	9·3323	11·097	HIP	BACK'G	300	194·0	TONGUE
7	10·888	12·947	PURLIN	PLUMB	300	229·0	TONGUE
8	12·443	14·796	PURLIN	EDGE	300	193·0	TONGUE
9	13·998	16·646	JACKS	EDGE	300	193·0	BLADE
10	15·554	18·495					

MOVE DECIMAL POINT 1, 2, OR 3 PLACES TO RIGHT AND UNITS BECOME TENS, HUNDREDS OR THOUSANDS. 0·4 & UNDER IGNORE. 0·5 & OVER ADD 1 TO UNITS.

DIFFERENCE IN JACKS
300 mm CENTRES 466·5 mm
350 mm CENTRES 544·5 mm
400 mm CENTRES 622 mm
450 mm CENTRES 700 mm
500 mm CENTRES 777·5 mm
550 mm CENTRES 855·5 mm
600 mm CENTRES 933 mm

$52\frac{1}{2}° = 391 \cdot 2$ mm RISE IN 300 mm RUN

RUN OF R	LENGTH OF RAFTER	LENGTH OF H&V	ROOF MEMBER AND BEVEL		FIGURES USED ON BLADE TONGUE		MARK CUT ON
1	1·6433	1·9229	R	PLUMB	300	391·0	TONGUE
2	3·2866	3·8458	R	SEAT	300	391·0	BLADE
3	4·9298	5·7686	H&V	PLUMB	424	391·0	TONGUE
4	6·5731	7·6915	H&V	SEAT	424	391·0	BLADE
5	8·2164	9·6144	H&V	EDGE	300	221·0	BLADE
6	9·8597	11·537	HIP	BACK'G	300	205·0	TONGUE
7	11·503	13·46	PURLIN	PLUMB	300	237·0	TONGUE
8	13·146	15·383	PURLIN	EDGE	300	183·0	TONGUE
9	14·789	17·306	JACKS	EDGE	300	183·0	BLADE
10	16·433	19·229					

MOVE DECIMAL POINT 1, 2, OR 3 PLACES TO RIGHT AND UNITS BECOME TENS, HUNDREDS OR THOUSANDS 0·4 & UNDER IGNORE. 0·5 & OVER ADD 1 TO UNITS.

DIFFERENCE IN JACKS
300 mm CENTRES 493 mm
350 mm CENTRES 575 mm
400 mm CENTRES 657 mm
450 mm CENTRES 739·5 mm
500 mm CENTRES 821·5 mm
550 mm CENTRES 904 mm
600 mm CENTRES 986 mm

55° = 427·5 mm RISE IN 300 mm RUN

RUN OF R	LENGTH OF RAFTER	LENGTH OF H&V	ROOF MEMBER AND BEVEL		FIGURES USED ON BLADE TONGUE		MARK CUT ON
1	1·7409	2·0076	R	PLUMB	300	427·5	TONGUE
2	3·4817	4·0153	R	SEAT	300	427·5	BLADE
3	5·2226	6·0229	H&V	PLUMB	424	427·5	TONGUE
4	6·9634	8·0305	H&V	SEAT	424	427·5	BLADE
5	8·7043	10·038	H&V	EDGE	300	212·0	BLADE
6	10·445	12·046	HIP	BACK'G	300	214·0	TONGUE
7	12·186	14·053	PURLIN	PLUMB	300	246·0	TONGUE
8	13·927	16·061	PURLIN	EDGE	300	173·0	TONGUE
9	15·668	18·069	JACKS	EDGE	300	173·0	BLADE
10	17·409	20·076					

MOVE DECIMAL POINT 1, 2, OR 3 PLACES TO RIGHT AND UNITS BECOME TENS, HUNDREDS OR THOUSANDS 0·4 & UNDER IGNORE. 0·5 & OVER ADD 1 TO UNITS.

DIFFERENCE IN JACKS
300 mm CENTRES 522 mm
350 mm CENTRES 609 mm
400 mm CENTRES 696 mm
450 mm CENTRES 783 mm
500 mm CENTRES 870 mm
550 mm CENTRES 957·5 mm
600 mm CENTRES 1044·5 mm

| \multicolumn{7}{c}{$57\frac{1}{2}° = 470\cdot25$ mm RISE IN 300 mm RUN} |

RUN OF R	LENGTH OF RAFTER	LENGTH OF H&V	ROOF MEMBER AND BEVEL		FIGURES USED ON BLADE TONGUE		MARK CUT ON
1	1·8593	2·1112	R	PLUMB	300	470·0	TONGUE
2	3·7186	4·2223	R	SEAT	300	470·0	BLADE
3	5·5779	6·3335	H&V	PLUMB	424	470·0	TONGUE
4	7·4372	8·4446	H&V	SEAT	424	470·0	BLADE
5	9·2964	10·556	H&V	EDGE	300	201·0	BLADE
6	11·156	12·667	HIP	BACK'G	300	223·0	TONGUE
7	13·015	14·778	PURLIN	PLUMB	300	258·0	TONGUE
8	14·874	16·889	PURLIN	EDGE	300	162·0	TONGUE
9	16·734	19·001	JACKS	EDGE	300	162·0	BLADE
10	18·593	21·112					

MOVE DECIMAL POINT 1, 2, OR 3 PLACES TO RIGHT AND UNITS BECOME TENS, HUNDREDS OR THOUSANDS 0·4 & UNDER IGNORE. 0·5 & OVER ADD 1 TO UNITS.

DIFFERENCE IN JACKS
300 mm CENTRES 557·5 mm
350 mm CENTRES 650·5 mm
400 mm CENTRES 743·5 mm
450 mm CENTRES 836·5 mm
500 mm CENTRES 929·5 mm
550 mm CENTRES 1022·5 mm
600 mm CENTRES 1115·5 mm

60° = 518·25 mm RISE IN 300 mm RUN

RUN OF R	LENGTH OF RAFTER	LENGTH OF H&V	ROOF MEMBER AND BEVEL		FIGURES USED ON BLADE TONGUE		MARK CUT ON
1	1·9961	2·2372	R	PLUMB	300	518·0	TONGUE
2	3·9921	4·4745	R	SEAT	300	518·0	BLADE
3	5·9881	6·7117	H&V	PLUMB	424	518·0	TONGUE
4	7·9842	8·949	H&V	SEAT	424	518·0	BLADE
5	9·9802	11·186	H&V	EDGE	300	190·0	BLADE
6	11·976	13·423	HIP	BACK'G	300	232·0	TONGUE
7	13·972	15·661	PURLIN	PLUMB	300	260·0	TONGUE
8	15·968	17·898	PURLIN	EDGE	300	151·0	TONGUE
9	17·964	20·135	JACKS	EDGE	300	151·0	BLADE
10	19·961	22·372					

MOVE DECIMAL POINT 1, 2, OR 3 PLACES TO RIGHT AND UNITS BECOME TENS, HUNDREDS OR THOUSANDS 0·4 & UNDER IGNORE. 0·5 & OVER ADD 1 TO UNITS.

DIFFERENCE IN JACKS
300 mm CENTRES 599 mm
350 mm CENTRES 698·5 mm
400 mm CENTRES 798·5 mm
450 mm CENTRES 898 mm
500 mm CENTRES 998 mm
550 mm CENTRES 1097·5 mm
600 mm CENTRES 1197·5 mm

STOBART DAVIES books for carpenters & joiners

MODERN PRACTICAL STAIRBUILDING & HANDRAILING - George Ellis Out of print for over 60 years. Ellis' last great work is once again reprinted. 21 detailed chapters and fully illustrated throughout.

TREATISE ON STAIRBUILDING & HANDRAILING - W Mowat
Originally published in 1900, Mowat's book remains one of the most lucid, best illustrated and authoritative works on the subject.

SPINDLE MOULDER HANDBOOK - Eric Stephenson
Deals with all aspects of this essential woodworking machine. A full range of practical work is covered - moulding. rebating, jointing, etc.

MOULDINGS & TURNED WOODWORK OF THE 16th, 17th, & 18th CENTURIES - T Small & C Woodridge
A large format volume with exceptionally lear and detailed drawings of stairs. doors, panelling, fireplaces , etc.

CIRCULAR WORK IN CARPENTRY & JOINERY - G Collings
A new edition of this standard work giving clear. easily understood explanations of the entire field of curved work in carpentry and joinery.